Miss J. Macmilla

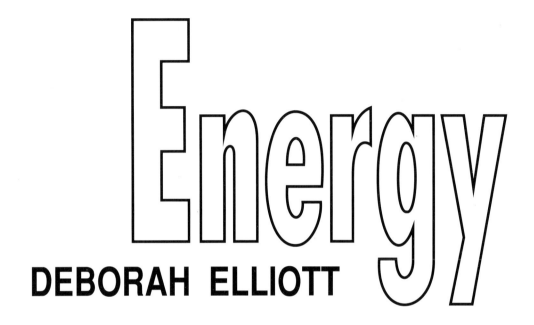

DEBORAH ELLIOTT

Wayland

Titles in the Into Europe series include:

Energy
Environment
Farming
Transport

Picture Acknowledgements:
British Petroleum (BP) 29; Chapel Studios 8 (top); Eye Ubiquitous 4 (top, Peter Blake), 7 (John Hulme), 12 (top), 39 (Mark Newman); Greenpeace 34 (Leitinger); Rex Features 11, 15 (bottom, Anastasselis); Science Photo Library 18, 24 (Simon Fraser) 36 (Martin Bond); Shell 15 (top); Tony Stone Worldwide 3, 8 (bottom), 23 (JW Sherriff), 25; Topham Picture Library 6, 16, 17, 20 (bottom), 22, 35; Wayland Picture Library 12 (bottom), 19, 20 (top), 29; Zefa 4 (bottom, Jed Sharpe), 10, 14 (bottom), 32, 37, 38, 41 (Starfoto), 43 (Eichorn - Z), 44 (Streichan); All artwork is by Malcolm Walker.

Designed by Malcolm Walker

Text based on *Energy in Europe* in the Europe series published in 1991.

First published in 1993 by Wayland (Publishers) Limited
61 Western Road, Hove, East Sussex BN3 1JD

© Copyright Wayland (Publishers) Limited

British Library Cataloguing in Publication Data
Elliott, Deb
　Energy. - (Into Europe Series)
　I. Title II. Series
　333.79

ISBN 0 7502 0756 6

Typeset by Kudos
Printed and bound by Canale & C.S.p.A. in Turin, Italy

Contents

What is energy 4
How we use energy 8
Western Europe 11
Eastern Europe 16
Getting fuel from the Earth 19
Coal – black gold? 21
Oil – liquid gold? 26
Gas – the energy of the future? 28
Nuclear energy – safe or sorry? 31
Saving energy 36
Other forms of energy 38
Energy tomorrow 44
Glossary 46
More information 47
Index 48

What is energy?

▲ We need energy to jump.

◀ Gas is used to heat this furnace.

▲ *The countries in Europe. Can you find where you live?*

Energy makes things move and work. All humans, plants and animals need energy to live. All machines need energy to work. We cannot actually see energy. However, we can see the results of energy being used all around us every day.

There are many different types of energy. All sources of energy come from the Sun.

We get our energy from food. It makes us breathe, walk and talk. Industries use energy in the form of electricity. Electricity is produced by burning coal, oil or gas. These are called fossil fuels and are found deep within the Earth's surface. Fossil fuels have been formed over thousands of years from the remains of dead plants and animals.

Fossil fuels provide energy for most of the industries in Europe. However, the day may soon come when the Earth will simply run out of fossil fuels.

Fossil fuels are not being made any more. There is only enough oil in the world to last another fifty years. Coal supplies will run out in about 300 years. Nuclear energy is made from a metal called uranium. This will run out in less than sixty years.

▲ *This poster shows an American lorry bringing energy supplies to Germany. Energy was needed to rebuild towns and factories in Europe after the Second World War ended in 1945.*

◀ *The Berlin Wall had divided East and West Germany since 1945. It was pulled down in October 1989 and Germany became one country again.*

Industries in the former East Germany now use less forms of energy that harm the environment.

Every hour of every day 1 million tonnes of fossil fuels are burned for energy in Europe. As they burn, fossil fuels give off smoke which contains poisonous gases. These gases are harmful to the environment.

Many people are trying to find other forms of energy which can be used instead of fossil fuels. There are lots of different forms of natural energy - from the power of the Sun, the wind, and the water in seas and oceans. These are called forms of alternative energy.

Europe uses more and more energy every day. New and safe forms of energy must be found.

How we use energy

▲ *Many people in Europe use coal to heat their homes. The coal is delivered to them each week. We get coal from mines under the ground.*

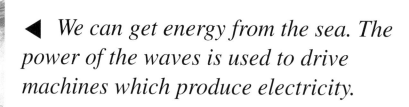

◀ *We can get energy from the sea. The power of the waves is used to drive machines which produce electricity.*

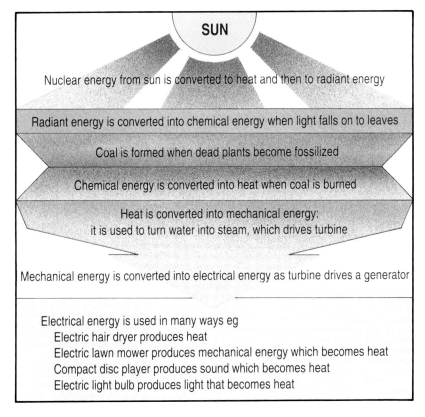

◀ *This diagram shows how all energy on Earth comes from the power of the Sun.*

▼ *The diagram below shows how we use energy in our homes.*

Gas, petrol, oil and electricity are popular forms of energy in homes and industries in Europe.

Electricity comes from burning fossil fuels and from nuclear power. We can also get electricity from alternative energy sources.

All electricity is produced in power-stations.

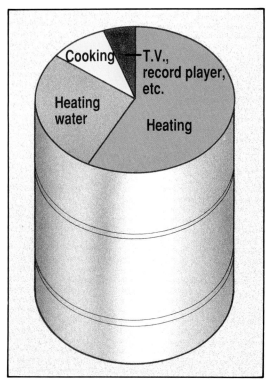

Energy is used in many ways and for all sorts of reasons. In Europe, over one-third of all energy is used in industry. The rest is used as fuel for cars, buses, trains and aeroplanes. It heats our homes, cooks our food and makes our televisions and hair dryers work.

◀ *Petrol is used to drive the engines in these cars. Petrol comes from oil.*

This diagram shows how much energy is produced and used by countries in the European Community (EC).

The EC is a group of twelve countries in Europe that have joined together to come up with ideas and plans for energy and farming. ▶

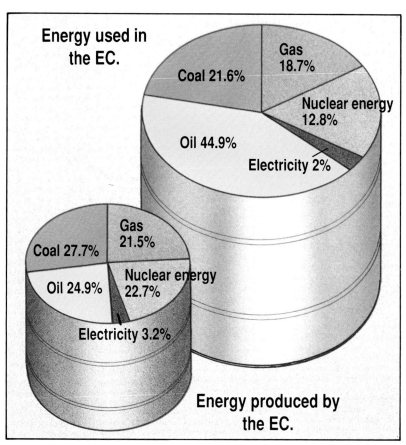

Western Europe

The man at the front of this photograph is Paul Henri Spaak of Belgium. He helped to set up the European Atomic Energy Commission (EURATOM) in 1958.

EURATOM was an organization of six countries that agreed to use and develop nuclear energy safely.

Oil is stored in these containers in Rotterdam in the Netherlands. ▶

▼ *Energy was needed to help rebuild cities in Europe after the Second World War ended in 1945.*

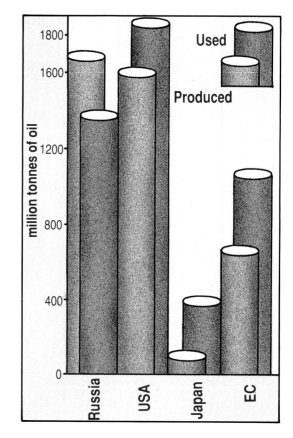

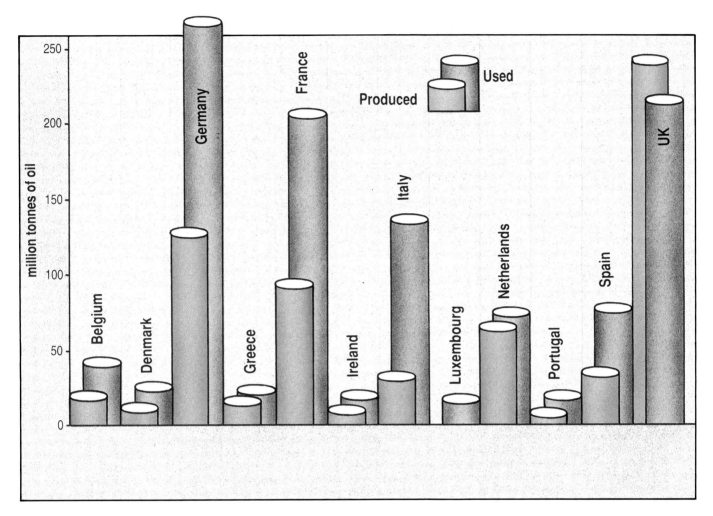

▲ *This diagram shows how much coal, oil and gas is used by many of the countries in Europe.*

◀ *The diagram opposite shows how much energy is produced and used by EC countries, Russia, Japan and the USA.*

Most oil is produced in the Middle East - countries such as Saudi Arabia, Iran and Kuwait. Europe also has oil fields in the North Sea. However, there is only enough oil left in the world to last about fifty years. Also, the price of oil in Europe is very high.

The rising price of oil has affected industry in western Europe badly. Because of this, many countries are using other forms of energy as well as oil, such as gas and nuclear energy. There has also been research into using alternative energy, such as solar energy (heat from the Sun).

The countries in the EC have come up with some policies on saving and re-using energy.

▼ *Tankers take oil from the sea to refineries, like this one in Rotterdam. Here the oil is passed through lots of machines. These turn the oil into petrol and fuel. The oil is carried all over Europe through pipelines that lie under the ground.*

▲ *This is one of the many oil platforms in the North Sea. Oil from the platforms is taken to refineries in Britain, Norway and the Netherlands.*

A meeting between members of EC countries. They are talking about ways to save energy. ▶

Eastern Europe

▲ *Queues of cars line up at a petrol station in Lithuania.*

The countries in eastern Europe buy all their oil and gas from Russia and Azerbaijan. These countries were both part of the USSR before it broke up in 1990. They produce one-fifth of the world's oil and one-third of the world's gas.

There have been many changes in eastern Europe in the 1990s. Before 1990, all the industry in eastern Europe was under the control of the government in each country. The coal-mines and factories were set targets as to how much they should produce.

These targets were not always reached. There was often not enough food or fuel in towns and cities.

Poland, Czechoslovakia and East Germany used to get energy from lignite. Lignite is a brown coal which gives off poisonous gases when burned. This has caused a great deal of pollution in these countries.

▲ *Cospa Mica in Romania is the most polluted town in Europe.*

◀ *Look at how much smoke is pouring out of this chimney in a power-station in Bitterfield in the east of Germany. The pollution is caused by lignite being burned to produce electricity.*

The changes to the governments in eastern Europe have led to changes in the way energy is produced and used. Many countries are now making sure that enough energy is made available for people to use.

Some western European industries have set up in eastern Europe. These industries use other sources of energy, such as gas, which do not cause as much pollution as lignite. This is safer and cleaner for the environment.

Getting fuel from the Earth

Coal, oil and gas are fossil fuels. They are found deep within the Earth. There are a number of ways to get fossil fuels from the Earth. In the past, miners used picks and shovels to dig coal from mines. Today, huge machines are used.

▼ *If the coal is not too deep under the ground, a machine is used to scrape away the earth. Miners can then dig out the coal. This is called opencast mining.*

These men are drilling for oil and gas in the North Sea. ▶

▼ *In 1988 there was a huge explosion on the Piper Alpha oil platform in the North Sea. Many people were injured or killed.*

Europe has its own oil fields in the North Sea. The oil belongs to Britain, the Netherlands, Germany and Norway.

Drilling for oil is a risky business, especially in the rough North Sea. Test drillings are carried out first. Once oil is found, it is carried ashore by pipelines that lie along the sea-bed.

Accidents on oil platforms can be very dangerous. Many lives can be lost, as with the Piper Alpha disaster.

Coal – black gold?

Coal was first discovered in Europe thousands of years ago. Since then it has heated homes and fuelled industries all over Europe.

Coal was one of the main reasons for the Industrial Revolution in Europe in the 1700s.

Before the Industrial Revolution, most people in Europe lived and worked on farms. Then, in 1769 a Scotsman called James Watt discovered that when coal was burned it gave off steam. This steam could be used to drive steam-engines.

Steam-engines were used in iron factories and cotton mills. Coal was also used to fuel steam trains and steam ships.

Industries were set up all over Europe. People left farming and went to work in factories. New towns grew up in the areas where coal was discovered.

Wales

Wales is famous for producing coal. Many towns and communities in the country grew up alongside coal-mines. In 1900 more coal was produced in Wales than anywhere else in the world. Most of this coal came from mines in the Rhondda Valley. There were fifty-three coal-mines in the Valley and most of the people living there worked in the coal industry.

Today the story is very different. There are only six coal-mines in use in the whole of Wales. Most stand empty and idle, like the one in the photograph. It is a very sad sight.

The coal industry in Wales and the rest of Britain suffered when people began to use coal brought in from the rest of Europe. It cost so much to get coal from British mines that it was cheaper to buy it from other countries.

Mines were closed down and miners lost their jobs. People left the mining towns to find work elsewhere. Communities and even families were broken up.

This unusual photograph was taken in the early evening. It shows the winding gear in a coal-mine. This is used to work the lifts that carry miners up and down to the mines. ▶

Coal was once the most important industry in Europe. The noise of the winding gear could be heard from dawn to dusk. Miners worked long hours in the deep, dark mines. Today things are different. Coal-mines stand empty and idle and miners have lost their jobs.

◀ *These spruce trees in Czechoslovakia have been killed by acid rain. Whole forests in the country have been destroyed.*

Coal is one of the major causes of pollution in the world. Coal gives off smoke and ash when it is burned. These contain a gas called carbon dioxide. The Earth is warmed by heat from the Sun. This heat escapes slowly into space. However, some gases, such as carbon dioxide, do not let all the heat escape. They trap heat like glass in a greenhouse. So the Earth is getting warmer. This is called the greenhouse effect.

That's not all. Smoke from power-staions and factory chimneys contains a chemical called sulphur. This pollutes moisture in the air. The polluted moisture falls to the ground as rain or snow, called acid rain.

Acid rain can kill plants and forests and destroy buildings.

The countries in eastern Europe where coal is the major source of energy are suffering from the effects of pollution and acid rain.

The coal industry has tried to find ways to stop pollution. In some countries in Europe, filters have been put into the chimneys in power-stations. These filters take out most of the sulphur that causes acid rain.

▲ *Coal is still the major source of energy in eastern Europe.*

Oil - liquid gold?

Oil is not only a source of energy. Petrol, plastics, glues, detergents, drugs, dyes and polishes can be made from crude oil.

Crude oil is also known as petroleum. It is a dark, sticky liquid found under the sea or in oil wells deep in the Earth. It is often found with natural gas.

Crude oil contains many different chemicals. These are separated from each other at a refinery (see page 14).

In the 1960s, the countries of Europe bought oil cheaply from the Middle East. However, during the 1970s there were political problems in the Middle East. For example, in 1973 war broke out between Israel and Egypt.

Less oil was available as a result. This caused the price to rise. In 1973, for instance, the price of oil went up by 400 per cent.

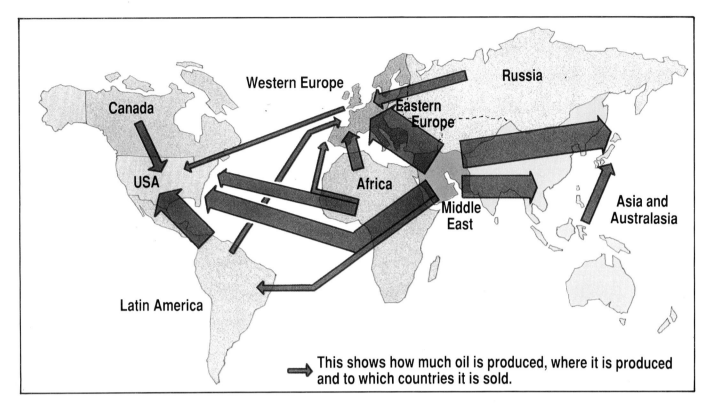

▲ *This diagram shows the countries of the world which produce oil. The arrows show which countries buy the oil. The thickness of the arrows tells us how much or how little oil is produced and sold.*

People tried to cut down the amount of oil and petrol they used. This meant that there was less need for oil. The rising price of oil led to problems for industries in Europe. Many could not afford to employ so many people. Some industries even went out of business.

The discovery of oil in the North Sea helped Europe. By 1987 the demand for energy was growing. Today, oil is the most important source of energy once again.

Gas - the energy of the future?

We use gas to heat our homes and to cook our food. Industries also use gas, especially the baking and glass-making industries.

Many people believe that one day gas will replace oil as the most important form of energy in Europe. Oil and coal supplies are running out. There are gas reserves all over the world that have not been touched.

Gas is also safer and cleaner for the environment than either oil or coal. It does not contain the poisonous chemical (carbon dioxide) that causes the greenhouse effect.

Countries in eastern Europe are starting to use gas instead of the brown coal, lignite, for energy. Gas fumes are less harmful and cause much less pollution.

Britain, the Netherlands, Russia, Norway and Algeria in northern Africa supply gas to the countries of Europe. The gas is carried from one country to another by trucks, tankers and pipelines. The map on page 30 shows how many pipelines run all over Europe.

Gas is also being used in power-stations to produce electricity. When gas is burned it gives off gases. These are used to turn a turbine. A turbine is an engine which forces wheels to go round and round to make electricity.

This hotel in the Algarve in the south of Portugal uses gas for heating and cooking. Lorries bring large amounts of gas to the hotel.

The gas is stored in these huge containers. ▶

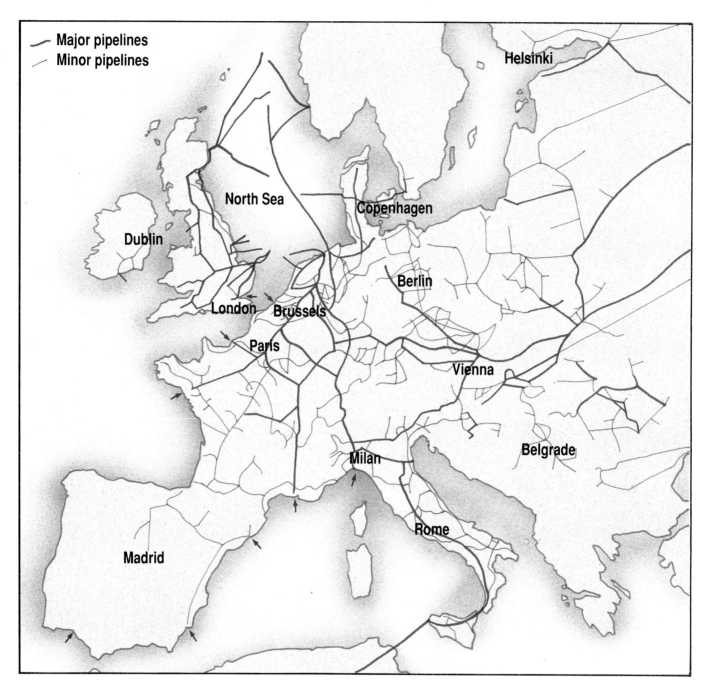

▲ *This map shows all the pipelines that carry gas around Europe. The small red arrows show the pipelines that lie along the sea-bed. Gas is also carried by trucks and tankers.*

Nuclear energy - safe or sorry?

We get nuclear energy from splitting atoms of the metal uranium. Atoms are the very tiny particles which all things are made up of. Everything we can see is made from millions of atoms. But the atoms are too small to be seen.

When uranium atoms are split, nuclear energy is given off. This is called a nuclear reaction. The atoms are split in a machine called a reactor.

Nuclear energy is used to produce electricity in nuclear power-stations. It works in the same way as coal, oil and gas. The nuclear energy heats water to make steam. The steam drives turbines which make electricity.

Many people think that nuclear energy is an alternative energy. They believe that it is safer and cleaner for the environment than coal, oil or gas.

◀ *This is the town of Schweinfurth in Germany.*

In the background stands a nuclear power-station. This produces electricity for the town.

However, nuclear energy produces radiation. This can be very dangerous if it is not treated properly. Waste from nuclear power-stations can harm people, animals and the environment. It is important that nuclear waste is put somewhere safe.

People who work in nuclear power-stations wear special clothes. These protect them from radiation.

You have probably heard of the Chernobyl disaster. It was the world's worst nuclear disaster. Thirty-one people died and hundreds of thousands suffered from the effects of radiation.

The disaster happened on 25 April, 1986. There was an explosion at the nuclear power-station in the town of Chernobyl in the Ukraine. Huge clouds of radioactive smoke gushed out into the sky. The clouds spread all over Europe. Forests, rivers, lakes and animals were all affected.

▲ *This diagram shows how the radioactive clouds spread after the Chernobyl disaster.*

France

France uses more nuclear energy than any other country in Europe.

Coal was once the main source of energy in France. All the coal-mines were in the north-west of the country. Getting the coal to other parts of France was very expensive.

France was also badly affected when the price of oil went up in the 1970s. The government decided to use nuclear energy. Power-stations were built all over the country.

▲ *Greenpeace is an organization that tries to help the environment. People who belong to Greenpeace believe that we should not use nuclear energy. They think it is not safe.*

The photograph above shows a Greenpeace protest in Austria. People let go of hundreds of balloons. They were showing their anger at the use of nuclear energy.

France, Belgium and Hungary use nuclear energy to produce most of their electricity. Germany, Britain and Spain use small amounts of nuclear energy. The Netherlands and Italy use hardly any at all.

Denmark, Iceland, Ireland, Portugal, Greece and Norway do not use any nuclear energy.

Europe needs other forms of energy to replace coal, oil and gas. It seems that nuclear energy is not the answer.

This pig was born blind. Its mother had been affected by radiation from the explosion at the nuclear power-station in Chernobyl. ▶

Saving energy

▲ As well as looking for other forms of energy, we need to save and re-use energy.

We can save energy by covering our heating pipes and boilers with a thick material. This stops heat from escaping. It is called insulation.

There are a number of ways that energy is saved in Europe. Many houses, especially in cold countries in northern Europe, have windows with double- or triple-glazing. These have two or three layers of glass that stop heat from getting out.

Many countries in northern and eastern Europe have Combined Heat and Power (CHP) stations. Here, heat is used to produce electricity. The electricity is used in factories. Then, it is carried along pipes to heat homes and factories.

◀ *We can re-use energy too.*

We use oil every day in our cars, homes and factories. Some of the oil is left over or wasted. It can be taken to factories like this one in Belgium.

Here the oil is passed through machines which get rid of any dirt. The oil can then be used again.

37

Other forms of energy

Fossil fuels are running out. The gases that are given off by burning fossil fuels cause pollution and acid rain. We can save energy and re-use energy. But this is not enough. We need other forms of energy.

We can get electricity from the power of the wind and water. Heat from rocks deep inside the Earth can also be used to produce electricity. The same is true for plants we see all around us every day.

◀ *Power from the Sun can be used to heat water and buildings.*

Power from the Sun, wind and water are forms of alternative energy. They will not harm the environment or cause pollution. Nor will they ever run out.

This is a power-station in Almeria in Spain. The large panels soak up energy from the Sun to heat water. ▶

Energy from the Sun is called solar power. Solar cells are used to trap this energy. These are tiny bits of metal covered with fine wires. They make electricity when they are hit by sunlight.

Each cell makes only a small amount of electricity. So lots of them are needed to power telephone lines and water pumps. Solar cells are also used to power watches and calculators.

In hot countries, sunlight can be used to heat water in homes. Houses have panels on the roofs. These soak up heat energy from the Sun during the day. The heat warms up water in tanks or pipes that touch the panels.

Windmills have been used in Europe for hundreds of years. In the past, they were used to grind flour into corn. They also pumped water from wet, marshy land in the Netherlands and parts of eastern Britain.

Nowadays, modern wind machines are used to make electricity. These machines are called wind turbines. They were first used in Denmark in the 1890s.

There is a photograph of wind turbines on page 44. The machines have blades that go round and round when the wind blows. These turn turbines which produce electricity.

Many countries in Europe are planning to build wind turbines. Wind energy is free. It does not use any fuel and it does not cause pollution.

However, we cannot rely on the wind. It is not always strong and does not blow in the same place at the same time. It should be used along with other forms of energy.

The water in oceans, rivers and waterfalls is always moving. It is full of energy. This energy can be used to make electricity. We get water (hydroelectric) power from tides, waves and hydroelectric (HEP) power-stations.

◀ *This dam was built in the Austrian Alps to make a large lake. A dam is a wall built across a river to hold back water.*

The lake is used to make electricity. The water passes through pipes into a hydroelectric power-station. The force of the water turns turbines. These make electricity

Norway and Switzerland get most of their energy from hydroelectric power.

Sweden

Up until 1970, most of the energy in Sweden came from oil. When the price of oil went up, the country had to find other forms of energy.

Today, hydroelectric and nuclear power-stations provide one-third of Sweden's energy. However, the Swedish people have voted to stop using nuclear energy by the year 2010.

There is much concern in Sweden about pollution and the environment. Forests and lakes in the country have been destroyed by acid rain. Swedish people do not want to use fossil fuels for energy. These are the cause of acid rain.

The Swedish people would like to use more alternative energy.

We can also get energy from tides. A long wall, called a barrage, is built across the widest part of the river where it meets the sea.

When the tide is high, water flows over and is trapped behind the barrage. When the tide is low, the water is let out through turbines which make electricity.

When the wind blows across water, we get waves. Waves can be very big and powerful. They are full of energy that could be used to make electricity. Scientists are trying to come up with water-power machines to do this. The perfect machine has not yet been found.

This is Reykjavik. It is the capital city of Iceland.

All the houses, shops and factories in Reykjavik are heated by energy from the Earth.

There are many volcanoes, geysers and hot springs in Iceland. These provide enough heat and steam to make electricity. ▶

There is a lot of heat buried deep within the Earth. Some of this heat comes to the surface in volcanoes, geysers and hot springs. These produce four times more heat energy than we use.

However, this heat is spread all over the world. Iceland is one of the few countries with enough underground stores of heat to make electricity.

Energy tomorrow

There are many forms of alternative energy that could replace fossil fuels. However, there are still problems with using alternative energy. Governments in Europe need to spend time and money trying to sort them out.

The Sun, wind, water, plants and the Earth will not destroy our environment. They do not cause acid rain. If used together they could provide Europe with all the energy it needs.

◀ *These are wind turbines. They are part of a wind farm where many turbines are used to make electricity.*

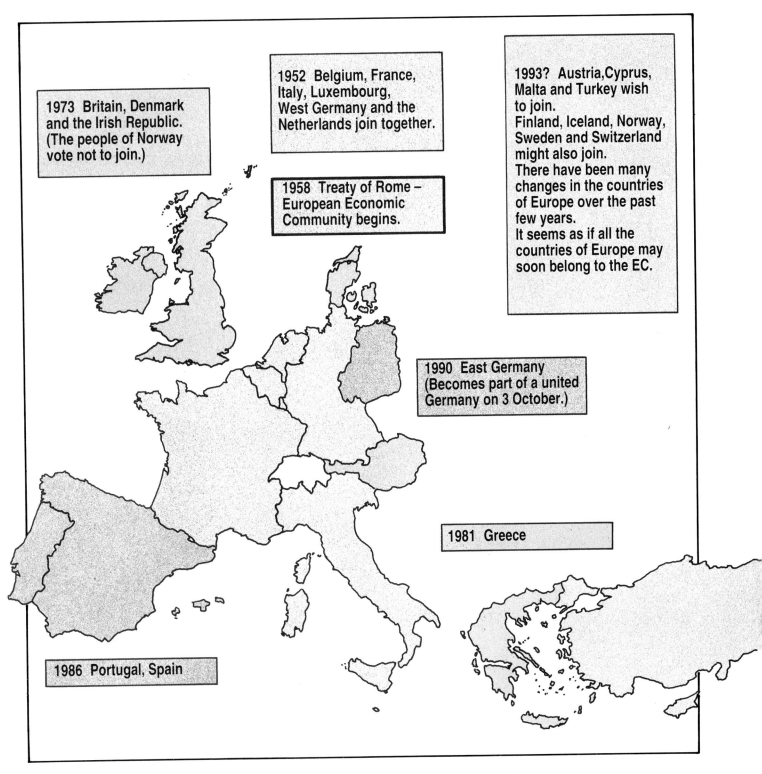

▲ *The countries of the EC and the year they joined.*

Glossary

Carbon dioxide A gas. We produce carbon dioxide when we breathe out and when we burn coal, oil and gas.

Communities Groups of people who live in one place.

Detergents Liquids that are used to clean things. They contain chemicals.

Environment The world around us. For example, animals, plants, rivers, mountains and the air we breathe.

Filters Machines that gases or liquids are passed through. The filters take out any dirt.

Geysers Springs under the ground which shoot out hot water and steam.

Hot springs Pools of very hot water that are found in certain countries, such as Iceland.

Industrial Revolution The time in Europe when people left farming to work in factories.

Natural All the things in the world that have not been made by people or machines.

Particles Very tiny bits of things, such as bits of dust.

Poisonous Harmful.

Pollution Things that spoil the environment.

Radiation Dangerous rays of energy that are produced by the fuel in nuclear power-stations.

Research To look closely into something to find out lots of information.

Sources The places where things begin.

Uranium A metal which is used to make nuclear energy.

Volcanoes Mountains which can explode and throw out hot ashes and burning rocks.

More information

Books to read

Energy and Power by Richard Spurgeon and Mike Flood (Usborne, 1990)
Power from Plants by Susan Bullen (Wayland, 1993)
Power from the Earth by Janet De Saulles (Wayland, 1993)
Power from the Sun by Susan Bullen (Wayland, 1993)
Power from Water by Hazel Songhurst (Wayland, 1993)
Power from Wind by Hazel Songhurst (Wayland, 1993)

Useful addresses

If you would like more information on energy in Europe, you can write to these organizations.

British Gas Education Service
PO Box 46
Hounslow
Middlesex TW4 6NP

British Nuclear Fuels PLC
Information Services
Sellafield
Seascale
Cumbria CA20 1BR

Conoco Ltd
Conoco Centre
Warwick Technology Park
Gallows Hill
Warwick CV34 6DA

Friends of the Earth
26-28 Underwood Street
London N1 7JQ

Index

Acid rain 24, 25, 38, 42, 44
Alternative energy 7, 9, 14, 31, 38-43, 44
Austria 34, 41
Azerbaijan 16

Belgium 12, 35, 37
Britain 12, 15, 20, 29, 35, 40

Chernobyl disaster 33, 35
Coal 6, 8, 12, 13, 19, 21-5, 28, 31, 34, 35
Coal-mining 8, 17, 19, 22, 23, 34
Combined Heat and Power (CHP) stations 37
Czechoslovakia 17, 24

Dams 41
Denmark 20, 35, 40

Earth, the 8, 19, 24, 26
 power from 38, 43, 44
Eastern Europe 16-18, 24, 25, 28
Electricity 6, 8, 9, 18, 29, 31, 32, 35, 37, 38, 39, 40, 41, 42, 43, 44
England 21
Environment, the 18, 24, 28, 32, 34, 38, 42, 44
EURATOM 11
European Community (EC), the 10, 11, 13, 14, 15, 45

Fossil fuels (coal, oil and gas) 6, 7, 9, 19, 38, 42, 44
France 12, 34, 35
Fuel 10, 14, 40

Gas 4, 6, 8, 13, 14, 16, 18, 19, 20, 26, 28-30, 31, 35
Germany 6, 7, 12, 17, 18, 20, 32, 34, 35
Geysers 43
Greece 35
'Greenhouse effect', the 24, 28
Greenpeace 34

Hot springs 43
Hungary 35
Hydroelectric power 41

Iceland 35, 43
Industrial Revolution, the 21
Industry 6, 9, 10, 14, 17, 18, 21, 22, 27, 28
Insulation 36
Ireland 35
Italy 35

Lignite (brown coal) 17, 18, 28
Lithuania 16

Netherlands, the 12, 15, 20, 29, 35, 40
North Sea, the 12, 13, 15, 20, 27
Norway 15, 29, 35, 41
Nuclear energy 6, 9, 11, 14, 31-5, 42

Oil 6, 9, 10, 12, 13-4, 15, 16, 19, 20, 22, 26-7, 28, 31, 34, 35, 37, 42

Petrol 9, 10, 14, 27

Pipelines 14, 20, 29, 30
Piper Alpha disaster 20
Plants, power from 38, 44
Poland 17
Pollution 17, 18, 24, 25, 28, 38, 40, 42
Portugal 29, 35
Power-stations 9, 18, 24, 25, 29, 39
 hydroelectric 41, 42
 nuclear 31, 32, 33, 34, 35, 42

Radiation 32, 33, 35
Refineries (oil) 14, 15, 26
Re-using energy 14, 36, 37, 38
Romania 17
Russia 12, 13, 16, 29, 33

Saving energy 14, 15, 36-7, 38
Spain 12, 35, 39
Sun, the 6, 8, 24
 power from 7, 38, 39, 44
Sweden 42
Switzerland 41

Tankers 12, 14, 29, 30
Turbines 29, 31, 40, 41, 44

Volcanoes 43

Wales 22
Water, power from 7, 8, 35, 41, 42, 44
Windmills 40
Wind, power from 7, 38, 40, 44